インプレス R&D [NextPublishing] 技術の泉 SERIES
E-Book / Print Book

TypeScriptで作る
シングルページアプリケーション

鈴木 潤 | 著

TypeScriptで
「型安全」な
JavaScript開発！

目次

はじめに―なぜTypeScriptなのか ··· 4
  本書の構成 ·· 4
  検証環境 ·· 4
  謝辞 ··· 5
  表記関係について ··· 5
  免責事項 ·· 5
  底本について ··· 5

第1章 TypeScriptとは ·· 6
  1.1 インタフェース ··· 6
  1.2 型アノテーション ··· 7
  1.3 アクセス修飾子 ··· 8
  1.4 構造的部分型 ··· 9
  1.5 列挙型 ··· 12

第2章 環境構築からHello Worldまで ··· 13
  2.1 Node.jsのインストール ·· 13
  2.2 TypeScriptのインストール ··· 14
  2.3 エディタについて ·· 15
  2.4 Hello World ·· 15
    2.4.1 ワーキングディレクトリの作成 ··· 15
    2.4.2 helloworld.tsの作成 ·· 15
    2.4.3 トランスパイル ··· 15
    2.4.4 実行 ··· 16

第3章 シングルページアプリケーションの作成 ··· 17
  3.1 データベースの準備 ·· 19
  3.2 サーバーサイドの開発 ·· 20
    3.2.1 tsconfig.jsonの作成 ··· 21
    3.2.2 package.jsonの作成とモジュールのインストール ·· 22
    3.2.3 実装 ··· 23
    3.2.4 ルーティング ·· 27
    3.2.5 　モデルの作成 ·· 30
    3.2.6 エラーページ用テンプレートの作成 ··· 31
    3.2.7 Daoクラス ·· 31
    3.2.8 サービスクラス ·· 36

3.2.9 コントローラクラス ………………………………………………………………	40

3.3 フロントエンドの開発 …………………………………………………………… 43

3.3.1 Angular-CLI のインストール ……………………………………………………	44
3.3.2 プロジェクトの作成 ………………………………………………………………	44
3.3.3 その他のライブラリのインストール …………………………………………	45
3.3.4 .angular-cli.json の修正 …………………………………………………………	45
3.3.5 プロキシの設定 ……………………………………………………………………	45
3.3.6 開発用タスクの実行 ………………………………………………………………	46
3.3.7 app-module の修正 ………………………………………………………………	47
3.3.8 モデルの作成 ………………………………………………………………………	50
3.3.9 グローバル CSS の設定 …………………………………………………………	50
3.3.10 index.html の修正 ………………………………………………………………	52
3.3.11 コンポーネントの追加 …………………………………………………………	53
3.3.12 サービスクラスの作成 …………………………………………………………	66
3.3.13 Store の作成 ……………………………………………………………………	71
3.3.14 Resolver クラスの作成 …………………………………………………………	73

3.4 スクリプト実行手順 ……………………………………………………………… 74

3.4.1 開発用スクリプトの実行手順 ……………………………………………………	74
3.4.2 本番環境用のスクリプト実行手順 ………………………………………………	75

著者紹介 ………………………………………………………………………………… 77

はじめに—なぜTypeScriptなのか

TypeScriptはMicrosoftが開発している、JavaScriptを拡張した静的型付けプログラミング言語です。TypeScriptで書かれたソースコードは、そのままではブラウザやNode.jsで実行できません。そのため、TypeScriptから、JavaScriptファイルを生成（トランスパイル）します。

ではなぜトランスパイルという工程を増やしてまで、TypeScriptを使うのでしょうか。筆者が一番大きなメリットだと感じているのは、多くの簡単なミス（例えば、メソッド名の打ち間違いや引数にbooleanを渡すべき箇所でnumberを渡している場合、あるいはそもそも構文のエラーが起きている場合など）を、トランスパイルの時点で（エディタがTypeScriptに対応していればリアルタイムに）検出できるようになることです。

素のJavaScriptであれば、このようなミスは実際にプログラムを動かしてみるまで分かりません。大抵の場合、実行時のエラーよりもトランスパイル時のエラーを楽に見つけることができるので、開発において非常に大きなメリットです。そして、これは開発の規模が大きくなるほど恩恵を受ける機会が多くなるでしょう。

本書を通じて、TypeScriptの利点と活用方法について触れてみてください。

本書の構成

第1～第2章で、代表的なTypeScriptの機能と、環境構築の手順を確認します。第3章では、実際にTypeScriptを使ってシングルページアプリケーションを作成します。

なお、本文中のソースコードは以下のリポジトリで公開しています。

URL:https://github.com/jsuzuki20120311/spa-typescript

検証環境

本文中のソースコードは以下の2台のマシンで動作検証を行っています。

OS: mac OS Sierra バージョン 10.12.6
MacBook Pro (13-inch, Mid 2010)
プロセッサ: 2.66 GHz Intel Core 2 Duo
メモリ 8 GB

OS: ubuntu 16.04 LTS 64 ビット

プロセッサ: Intel Core i5-3230M

メモリ 4 GB

また、Node.js と npm は、いずれのマシンでも以下のバージョンを使用しています。

Node.js: v8.9.4

npm: 5.6.0

謝辞

本書の執筆にあたり私を支えてくれた全ての人に感謝します。それから、私が TypeScript を使ってプログラミングをするきっかけを与えてくれた、今はクローズしてしまった、とあるプロジェクトに感謝します。

表記関係について

本書に記載されている会社名、製品名などは、一般に各社の登録商標または商標、商品名です。会社名、製品名については、本文中では ©、®、™ マークなどは表示していません。

免責事項

本書に記載された内容は、情報の提供のみを目的としています。したがって、本書を用いた開発、製作、運用は、必ずご自身の責任と判断によって行ってください。これらの情報による開発、製作、運用の結果について、著者はいかなる責任も負いません。

底本について

本書籍は、技術系同人誌即売会「技術書典」で頒布されたものを底本としています。

第1章 TypeScriptとは

　実際にプログラミングを始める段階で、使用する言語のすべてを把握する必要は無いでしょう。しかし、その代表的な機能を知っておくことは、学習を進めるうえで間違いなく助けになります。

　TypeScriptの言語仕様は、EcmaScript5の仕様に以下の3つの機能を追加したものです。

・EcmaScript2015由来の機能
　　―class 構文
　　―let, const
　　―ラムダ式
　　etc…
・EcmaScript2016由来の機能
　　―デコレータ
　　―async/await構文
・**TypeScript独自の機能**
　　―インタフェース
　　―型アノテーション
　　―アクセス修飾子
　　―列挙型
　　etc…
　本章では、3番目のTypeScript独自の機能を中心に確認していきます。

1.1 インタフェース

　EcmaScript5やEcmaScript2015に存在しない機能として、**インタフェース**があります。インタフェースにはプロパティとメソッドの定義のみができ、実装することはできません。たとえば、以下のRunnableインタフェースは string型のnameプロパティと、booleanを返すrunメソッドが定義されています。

リスト1.1.1 Runnable.ts

```
interface Runnable {
  name: string;
  run(): boolean;
```

```
}
```

上記のRunnableインタフェースを実装したTaskクラスを作る場合、以下のように**implements キーワード**を使用します。

implementsキーワードを使用しインタフェースを指定することで、そのクラスにインタフェースで定義したプロパティとメソッドの実装を強制させることができます。インタフェースで定義したプロパティとメソッドがどれか1つでも実装されていない場合、トランスパイルでエラーが発生します。

リスト1.1.2 Task.ts

```
class Task implements Runnable {

  name: string

  constructor(name: string) {
    this.name = name;
  }

  run(): boolean {
    let result = false;
    // do something
    return result;
  }
}
```

1.2 型アノテーション

TypeScriptでは、**型アノテーション**を使用して、変数に静的に型を指定することができます。以下のように型アノテーションは変数名の後に指定し、間に :(コロン) をつけます。

リスト1.2.1

```
let hoge: number = 1;
```

また、クラスのプロパティに指定する場合、リスト1.2.2の3行目のようにします。

リスト1.2.2

```
class Hoge {
```

第1章 TypeScriptとは 7

```
  private fuga: string;

  constructor(fuga: string) {
    this.fuga = fuga;
  }
}
```

1.3 アクセス修飾子

TypeScriptでは、public、protected、privateの**アクセス修飾子**が使用できます。なお、何も指定しない場合、publicとなります。

・public
　—外部のクラスからもアクセス可能。
・protected
　—クラス内とそのクラスのサブクラスからのみアクセスが可能。
・private
　—クラス内からのみアクセスが可能。

また、TypeScriptのバージョン2以降では、以下のようにコンストラクタにprivateやprotectedが指定できるようになりました。これにより、より安全にSigletonパターンが使用できるようになりました。

リスト1.3 コンストラクタにprivate修飾子を指定したSingletonクラス

```
class Singleton {

  private static instance: Singleton;

  private constructor() {
    // do nothing
  }

  public static getInstance(): Singleton {
    if (!Singleton.instance) {
      Singleton.instance = new Singleton();
    }
    return Singleton.instance;
  }
```

```
}
```

なお、インタフェースで定義したプロパティとメソッドは、自動的にpublicとなります。実装しているクラスでprotectedやprivateを指定すると、トランスパイルエラーが発生します。

1.4 構造的部分型

TypeScriptには、JavaやC#のようなクラスベースのオブジェクト指向プログラミング言語とは異なる考え方に、**構造的部分型**があります。

構造的部分型とは、型の派生関係をextendsやimplementsからではなく、オブジェクトの構造から判断する仕組みのことです。

例えばJavaにおいて以下のような異なる2つの、但し構造は同じインタフェースがあり、それぞれに実装したクラスがあったとします。

リスト1.4.1 Flyable.java

```java
public interface Flyable {
  void fly();
}
```

リスト1.4.1 Dragon.java

```java
public interface Dragon {
  void fly();
}
```

リスト1.4.2 Raven.java

```java
public class Raven implements Flyable {
  public void fly() {
    System.out.println("fly");
  }
}
```

リスト1.4.3 BlueDragon.java

```java
public class BlueDragon implements Dragon {
  public void fly() {
    System.out.println("fly");
  }
}
```

その上で、リスト1.4.4の処理を書いた場合、9行目でコンパイルエラーが発生します。

第1章 TypeScriptとは　9

リスト 1.4.4 Main.java

```java
public class Main {

    public static void main(String[] args) {
        Main main = new Main();

        Flyable raven = new Raven();
        Dragon blueDragon = new BlueDragon();
        main.register(raven);
        main.register(blueDragon);
    }

    private void register(Flyable flyable) {
        // do something
    }
}
```

これは、BlueDragon クラスが Flyable インタフェースを実装しておらず、Java が構造ではなく extends や implements キーワードを元に型の派生関係を判断しているためです。

一方、TypeScript で似たようなコードを書いた場合、トランスパイルエラーが発生しません。

リスト 1.4.5 Flyable.ts

```typescript
interface Flyable {
    fly(): void;
}

export default Flyable;
```

リスト 1.4.6 Dragon.ts

```typescript
interface Dragon {
    fly(): void;
}

export default Dragon;
```

リスト 1.4.7 Raven.ts

```typescript
import Flyable from './Flyable';

class Raven implements Flyable {
```

10 第 1 章 TypeScript とは

```
  fly(): void {
    console.log('fly');
  }
}

export default Raven;
```

リスト1.4.8 BlueDragon.ts

```
import Dragon from './Dragon';

class BlueDragon implements Dragon {

  fly(): void {
    console.log('fly');
  }
}

export default BlueDragon;
```

　Javaのケースでコンパイルエラーが起きた行に相当する17行目があっても、TypeScriptではトランスパイルエラーは発生しません。これは構造が同じであるためです。

リスト1.4.9 Main.ts

```
import BlueDragon from './models/BlueDragon';
import Raven from './models/Raven';
import Flyable from './models/Flyable';

class Main {

  register(flyable: Flyable): void {
    flyable.fly();
  }
}

let main = new Main();

let raven = new Raven();
let blueDragon = new BlueDragon();
main.register(raven);
main.register(blueDragon);
```

1.5 列挙型

TypeScript では**列挙型（enum）**がサポートされています。列挙型は以下のように作成できます。

リスト1.5.1

```
enum Animals{
  Cat,
  Dog,
  Pig
}

console.log(Animals.Cat, Animals.Dog, Animals.Pig); // 出力結果: 0 1 2
```

以下のように明示的に数値を指定できます。

リスト1.5.2

```
 enum Animals{
  Cat = 100,
  Dog = 200,
  Pig = 300
}

console.log(Animals.Cat, Animals.Dog, Animals.Pig); // 出力結果: 100
200 300
```

第2章 環境構築からHello Worldまで

‖‖
　本章では環境構築を行います。2章の内容に関しては、自身の端末のインストール済みプログラムや環境変数などと比較しながら適宜読み替えてください。
‖‖

2.1 Node.jsのインストール

　まずNode.jsをインストールします。Node.jsはサーバーサイドのJavaScript実行環境です。本書ではNode.jsを使用してTypescriptからJavaScriptへのトランスパイルやサーバーサイドのプログラムを実行します。

　Node.jsのインストールにはnodebrewの使用をおすすめします。nodebrewはNode.jsのバージョン管理ツールです。Node.jsはバージョンアップの頻度が高く、バージョンの変更の度にアンインストールとインストール作業を繰り返すのは非常に手間がかかります。nodebrewを使用することで、この手間を削減できます。

　ターミナルでリスト2.1.1のコマンドを実行しnodebrewをインストールします。

リスト2.1.1

```
$ curl -L git.io/nodebrew | perl - setup
```

　nodebrewのインストールが完了したら、nodebrewのパスを通します。bashを使用している場合、「.bashrc」にファイルに以下のコードを追記します。

リスト2.1.2

```
export PATH=$HOME/.nodebrew/current/bin:$PATH
```

　nodebrewのパスを通したら、nodebrewを使用してNode.jsをインストールします。今回は執筆時点でLTSバージョンである8.9.4をインストールします。

リスト2.1.3

```
$ nodebrew install-binary v8.9.4
```

正しくNode.jsのv8.9.4がインストールされたかどうか確認します。

リスト2.1.4

```
$ nodebrew list
```

これはnodebrewからインストールされたNode.jsのバージョンを一覧表示するコマンドです。インストールが完了していれば、v8.9.4が表示されます。

次に、使用するNode.jsのバージョンを指定します。今回はv8.9.4を使用するので、以下のコマンドを実行します。

リスト2.1.5

```
$ nodebrew use v8.9.4
```

Node.jsのv8.9.4を使用するよう指定されたことを、以下のコマンドで確認します。

リスト2.1.6

```
$ node --version
```

正しく指定されていれば『v8.9.4』と表示されます。

2.2 TypeScriptのインストール

TypeScriptをインストールします。TypeScriptのインストールにはnpmコマンドを使用します。npmはNode.jsと共にインストールされるモジュール管理ツールです。

リスト2.2.1

```
$ npm install typescript --global
```

明示的にTypeScriptのバージョンを指定する場合はリスト2.2.2のように@以降にバージョン番号を指定します。（本書中のソースコードは、全て2.4.2を使用して動作確認をしています。）

リスト2.2.2

```
$ npm install typescript@2.4.2 --global
```

リスト2.2.3のコマンドでTypeScriptのバージョンを表示し、インストールが正常に完了した

か確認します。

リスト 2.2.3

```
$ tsc --version
```

2.3 エディタについて

エディタについては使い慣れたエディタがあればそれを使用するのが一番ですが、特にこだわりが無い場合、あるいは使い慣れたエディタがTypeScriptに対応していない場合はVisualStudioCodeの使用をおすすめします。

VisualStudioCodeはTypeScript開発元と同じMicrosoft製のエディタです。Microsoft製だけあり、特にプラグインを導入することなくTypeScriptに対応しており、Windows版だけでなくMac版、Linux版もリリースされています。

VisualStudioCodeは、以下のURLからダウンロードできます。

URL:https://code.visualstudio.com/download

2.4 Hello World

新たにプログラミング言語を学ぶ時や、プログラミングをはじめる環境が整ったことを確認する際には、伝統的に『Hello World』という文字列を表示するプログラムを作成します。本書もそれに習い、ハローワールドプログラムを作ります。

2.4.1 ワーキングディレクトリの作成

適当な場所にワーキングディレクトリを作成します。今回はディレクトリ名を『chapter-2』とします。ディレクトリを作成したら、『helloworld.ts』ファイルを作成します。

2.4.2 helloworld.ts の作成

作成した『helloworld.ts』ファイルに次のコードを書き保存します。

```
console.log('Hello World');
```

2.4.3 トランスパイル

コードが書けたら、TypeScriptからJavaScriptへの変換を行います。この変換はコンパイルやトランスパイル、あるいはトランスコンパイルなどと呼ばれます。本書ではトランスパイル

第2章 環境構築から Hello World まで   15

に統一します。

作成した『chapter-2』ディレクトリに移動し、次のコマンドを実行します。

```
$ tsc helloworld.ts
```

正常に完了したら、ディレクトリ内に『helloworld.js』ファイルが作成されていることを確認します。この『helloworld.js』がトランスパイルされた結果であり、実際に実行されるコードが書かれたファイルです。

2.4.4 実行

正常にトランスパイルが完了したら、実際に実行します。実行するには次のコマンドを実行します。実行するのはトランスパイル後のファイルなので、指定するのは『helloworld.ts』ではなく『helloworld.js』である点に注意します。

```
$ node helloworld.js
```

ターミナルに『Hello World』が表示されたら成功です。

第3章 シングルページアプリケーションの作成

第3章では以下の機能を持った記事管理シングルページアプリケーションを作成します。

・記事の投稿
・記事の閲覧
・記事の更新
・記事の削除

サーバーサイド、フロントエンドそれぞれに以下のフレームワークを使用します。
・Express
—Node.js用Webアプリケーションフレームワークです。
・Angular
—Googleが開発を主導しているフロントエンドWebアプリケーションフレームワークです。

　本章ではワーキングディレクトリとして『chapter-3』という名前のディレクトリ下で作業する前提で進めます。さらに、『chapter-3』ディレクトリの下に『client』と『server』の2つのディレクトリを作成します。これはそれぞれ、フロント側のプロジェクト、サーバー側のプロジェクト用のディレクトリとなります。

```
chapter-3/
   ├ client/    フロント側プロジェクト用ディレクトリ
   └ server/    サーバー側プロジェクト用ディレクトリ
```

　図3.1〜3.4は完成形の画面イメージです。

図3.1 画面イメージ 記事一覧画面

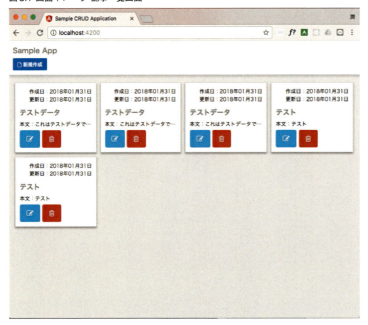

図3.2 画面イメージ 記事 内容確認画面

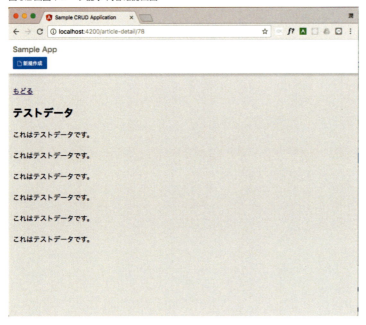

図3.3画面イメージ記事 記事登録画面

図3.4画面イメージ記事　記事更新画面

3.1 データベースの準備

　本書ではMySql5.7の使用を前提に進めます。また、MySQLのインストール手順や設定手順

は割愛します。MySQLのインストールが完了したら、リスト3.1のSQL文を実行しarticleテーブルを作成します。

リスト3.1

```
create table article
(
id int not null auto_increment,
title varchar(2000) not null,
body longtext null,
created_at datetime null,
updated_at datetime null,
constraint 'PRIMARY'
primary key (id),
constraint article_id_uindex
unique (id)
);
```

3.2 サーバーサイドの開発

サーバーサイドのアプリケーションは以下の3階層の構成で、APIを作成します。

・コントローラクラス

　—個々のリクエストに対してレスポンスを返すクラスです。今回の場合、内部でサービスクラスのインスタンスを保持し、そのインスタンスの処理結果でもってレスポンスの内容を操作します。

・サービスクラス

　—実処理を行います。内部でDaoのインスタンスを作成し、そのインスタンスを通してDBとのやりとりや行います。

・Daoクラス

　—データベースへのアクセスを行います。

サーバーサイドの開発を進めるにあたって、『server』ディレクトリ直下の構成は以下のようにします。ディレクトリに関しては最初の段階ですべて作ってしまうことをおすすめします。

```
chapter-3/
  └ server/     サーバー側プロジェクト用ディレクトリ
     ├ config/        設定保持系のクラス
     ├ controllers/   コントローラ用ディレクトリ
     ├ dao/           Daoクラス用ディレクトリ
     ├ json/          jsonファイル置き場
```

20　　第3章 シングルページアプリケーションの作成

```
├ models/              Modelクラス用ディレクトリ
├ node_modules/        npmモジュール用ディレクトリ
├ public/              静的ファイル置き場
├ routes/              ルーティング用ディレクトリ
├ services/            サービスクラス用ディレクトリ
├ views/               テンプレート用ディレクトリ
├ www/                 Webサーバー起動処理用ディレクトリ
├ app.ts               Expressアプリケーション設定処理
├ tsconfig.json
└ package.json
```

3.2.1 tsconfig.jsonの作成

まず『tsconfig.json』という名前のファイルをserverディレクトリに作成します。これはTypeScriptで記述したファイルをトランスパイルする際の、オプションなどを設定するファイルです。ファイルに記述する内容ですが、今回は以下の内容で作成します。

リスト3.2.1

```
{
  "compilerOptions": {
    "target": "ES5",
    "module": "commonjs",
    "moduleResolution": "node",
    "sourceMap": true,
    "emitDecoratorMetadata": true,
    "experimentalDecorators": true,
    "lib": ["es2015"],
    "noImplicitAny": true,
    "suppressImplicitAnyIndexErrors": true,
    "typeRoots": [
      "./node_modules/@types"
    ],
    "alwaysStrict": true
  },
  "compileOnSave": true,
  "exclude": [
    "node_modules"
  ]
}
```

3.2.2 package.jsonの作成とモジュールのインストール

次にpackage.jsonを設定します。通常、新規作成する場合はリスト3.2.2.1のコマンドを実行しますが、今回は予め必要なモジュールを指定したファイルを作成します。リスト3.2.2.1は実行せず、リスト3.2.2.2の内容でpackage.jsonという名前でファイルを新規作成します。

リスト3.2.2.1

```
$ npm init -y
```

リスト3.2.2.2 package.json

```
{
  "name": "sample-crud-application-server",
  "version": "1.0.0",
  "main": "./www/index.js",
  "scripts": {
    "start": "NODE_ENV=production node ./www/index.js",
    "build": "tsc",
    "dev": "NODE_ENV=development tsc && tsc --watch & node-dev
./www/index.js"
  },
  "license": "MIT",
  "dependencies": {
    "body-parser": "~1.15.2",
    "ejs": "~2.5.6",
    "express": "~4.14.0",
    "mysql": "~2.11.1"
  },
  "devDependencies": {
    "@types/body-parser": "~0.0.34",
    "@types/ejs": "~2.3.33",
    "@types/express": "~4.0.33",
    "@types/mysql": "~0.0.31",
    "node-dev": "~3.1.3",
    "typescript": "~2.4.2"
  }
}
```

package.jsonを作成したら、リスト3.2.2.3のコマンドを実行し、必要なnpmモジュールをインストールします。

リスト 3.2.2.3

```
$ npm install
```

リスト3.2.2.3のコマンドを実行すると、`package.json`のdependenciesとdevDependencies
で管理されているnpmモジュールがインストールされます。

3.2.3 実装

まずMySQLに接続するための設定を保持するためのファイルを作ります。今回は設定をjson
ディレクトリ内に`db_config.json`というファイル名でJSONファイルを作成します。ファイ
ルを作成したら以下のように、host、port、user、password、databaseに、自身のデータベー
スの接続情報を記述し保存します。

```
chapter-3/
  └ server/   サーバー側プロジェクト用ディレクトリ
      ├ config/
      │   └ db-config-manager.ts
    // 省略
      │
      ├ json/
      │   └ db_config.json // データベースへの接続設定ファイル
    // 省略
```

リスト 3.2.3.1 server/json/db_confoig.json

```json
{
  "host": "localhost",
  "port": 3306,
  "user": "root",
  "password": "hogehoge",
  "database": "sample_crud_app_db"
}
```

データベース接続情報のファイルを作成したら、次に接続情報の設定を保持するためのクラ
スを作成します。

リスト 3.2.3.2 server/config/db-config-manager.ts

```typescript
import * as mysql from 'mysql';
import * as fs from 'fs';

export class DbConfigManager {
```

第3章 シングルページアプリケーションの作成 | 23

```
  private static dbConfig: mysql.IConnectionConfig;

  static initialize(): void {
    const data = fs.readFileSync('./json/db_config.json', 'utf-8');
    DbConfigManager.dbConfig = JSON.parse(data);
  }

  static getConfig(): mysql.IConnectionConfig {
    if (!DbConfigManager.dbConfig) {
      throw new Error('Did not initialized!');
    }
    return DbConfigManager.dbConfig;
  }

}
```

　次にWebサーバー起動処理を書いていきます。『www』ディレクトリ内に、リスト3.1.3の内容でindex.tsファイルを作成します。このindex.ts内では、4行目でデータベース接続インスタンスの初期化も行っています。

```
chapter-3/
  └ server/  サーバー側プロジェクト用ディレクトリ
  // 省略
      ├ www/
      │   └ index.ts // Webサーバー起動処理
  // 省略
```

リスト3.2.3.3 server/www/index.ts

```
import * as http from 'http';
import app from '../app';
import { DbConfigManager } from '../config/DbConfigManager';

DbConfigManager.initialize();

const server = http.createServer(app);
const port = normalizePort(process.env.PORT || '3000');
server.listen(port);
server.on('error', onError);
server.on('listening', onListening);
```

```typescript
/**
 * Normalize a port into a number, string, or false.
 */
function normalizePort(val: string) {
  const port = parseInt(val, 10);

  if (isNaN(port)) {
    // named pipe
    return val;
  }

  if (port >= 0) {
    // port number
    return port;
  }

  return false;
}

/**
 * Event listener for HTTP server "error" event.
 */
function onError(error: any) {
  if (error.syscall !== 'listen') {
    throw error;
  }

  const bind = typeof port === 'string'
      ? 'Pipe ' + port
      : 'Port ' + port;

  // handle specific listen errors with friendly messages
  switch (error.code) {
    case 'EACCES':
      console.error(bind + ' requires elevated privileges');
      process.exit(1);
      break;
    case 'EADDRINUSE':
      console.error(bind + ' is already in use');
      process.exit(1);
      break;
    default:
```

```
    throw error;
  }
}


/**
 * Event listener for HTTP server "listening" event.
 */
function onListening() {
  const addr = server.address();
  const bind = typeof addr === 'string'
      ? 'pipe ' + addr
      : 'port ' + addr.port;
  console.log(addr + bind);
}
```

次に、リスト3.2.3.3の2行目で参照している app.ts をリスト3.1.3.4の内容で作成します。

```
chapter-3/
  ├ client/   フロント側プロジェクト用ディレクトリ
  └ server/   サーバー側プロジェクト用ディレクトリ
  // 省略
      └ app.ts
```

リスト 3.2.3.4 server/app.ts

```
import * as bodyParser from 'body-parser';
import 'ejs';
import * as express from 'express';
import * as path from 'path';
import api from './routes/api';
import index from './routes/index';


/**
 * Expressアプリケーションオブジェクト
 * @type {Express}
 */
const app = express();

// テンプレートエンジンに ejs を使用するための設定
app.set('views', __dirname + '/views');
app.set('view engine', 'ejs');
```

```typescript
app.use(bodyParser.urlencoded({ extended: true }));
app.use(bodyParser.json());

// 静的ファイルのルーティング
app.use(express.static(path.join(__dirname, 'public')));
app.use('/', index);
app.use('/api', api);

// 404のルーティング
app.use((req, res, next) => {
  const err = {
    status: 404,
    message: 'Not Found.'
  };
  next(err);
});

// エラーハンドラ
app.use((err: any, req: express.Request, res: express.Response, next:
express.NextFunction) => {
  err.status = err.status || 500;
  res.status(err.status);
  if (req.xhr) {
    res.send({ status: err.status, message: err.message });
  } else {
    res.render('error', { status: err.status, message: err.message
});
  }
});

export default app;
```

3.2.4 ルーティング

　app.tsの作成が完了したら次にルーティングの設定を行います。ここでは以下の機能を実装します。

・以下のURLでは常にpublicディレクトリ直下の『index.html』を返します。この設定はフロント側でのルーティングと対応させるための設定です。

　　—http://ドメイン:ポート番号/

—http://ドメイン:ポート番号/create-article/

—http://ドメイン:ポート番号/update-article/ + 記事データのID

—http://ドメイン:ポート番号/view-article/ + 記事データのID

・以下のURLで各APIを呼び出します。ソースコード上は、それぞれコントローラの対応す
るメソッドを呼び出しています。

　　　—http://ドメイン:ポート番号/api/v1/article/

　　　　　POSTメソッド

　　　　　　記事データ一覧を返す。

　　　—http://ドメイン:ポート番号/api/v1/article/ + 記事データのID + .json

　　　　　GETメソッド

　　　　　　記事データを返す

　　　　　PUTメソッド

　　　　　　記事データを更新する。

　　　　　DELTEメソッド

　　　　　　記事データを削除する。

以上の処理を、URLの階層構造と同じように以下の構成で、ディレクトリとTypeScriptファ
イルを作成します。ルーティングの設定にはExpress.Routerのget, post, put, delete などのメ
ソッドとuse メソッドを使用します。

```
chapter-3/
  └ server/   サーバー側プロジェクト用ディレクトリ
  // 省略
      ├ routes/
  │   ├ api/
  │   │   ├ v1/
  │   │   │   ├ article.ts
  │   │   │   └ index.ts
  │   │   └ index.ts
  │   └ index.ts
  // 省略
```

リスト3.2.4.1 server/routes/index.ts

```
import * as express from 'express';
import * as path from 'path';

const index = express.Router();
const sendIndexHtml = (req: express.Request, res: express.Response)
=> {
  res.sendFile(path.join(__dirname, '../', 'public', 'index.html'));
```

28 第3章 シングルページアプリケーションの作成

```
};
index.get('/', sendIndexHtml);
index.get('/createArticle', sendIndexHtml);
index.get('/updateArticle/:id', sendIndexHtml);
index.get('/viewArticle/:id', sendIndexHtml);

export default index;
```

リスト 3.2.4.2 server/routes/index.ts

```
import * as express from 'express';
import v1 from './v1';

const api = express.Router();
api.use('/v1', v1);

export default api;
```

リスト 3.1.4.3 server/routes/index.ts

```
import * as express from 'express';
import article from './article';

const v1 = express.Router();
v1.use('/article', article);

export default v1;
```

リスト 3.2.4.4 server/routes/index.ts

```
import * as express from 'express';
import { ArticleController } from
'../../../controllers/ArticleController';

const article = express.Router();

article.get('/', (req, res, next) => {
  const articleController = new ArticleController();
  articleController.index(req, res, next);
});

article.post('/', (req, res, next) => {
  const articleController = new ArticleController();
  articleController.create(req, res, next);
```

```
});

article.get('/all.json', (req, res, next) => {
  const articleController = new ArticleController();
  articleController.all(req, res, next);
});

article.get('/count.json', (req, res, next) => {
  const articleController = new ArticleController();
  articleController.count(req, res, next);
});

article.get('/:id\.json', (req, res, next) => {
  const articleController = new ArticleController();
  articleController.read(req, res, next);
});

article.put('/:id\.json', (req, res, next) => {
  const articleController = new ArticleController();
  articleController.update(req, res, next);
});

article.delete('/:id\.json', (req, res, next) => {
  const articleController = new ArticleController();
  articleController.delete(req, res, next);
});

export default article;
```

3.2.5　モデルの作成

　次はモデルを作成します。ここで言うモデルとは、データベースに格納される1つ1つのレコードの各カラムの属性を持つクラス、またはインタフェースのことを指します。所謂ビジネスロジックのことではなく、また、データベースへアクセスするための機能も持ちません。

　また、今回はサンプルアプリケーション全体で使用するAppErrorクラスもmodelsディレクトリに作成してしまいます。他にもエラー系のクラスが増えるようであればerrorsやexceptionsなどの名前でディレクトリを作成し、そこに置くべきでしょう。

リスト3.2.5.1 server/models/article.ts

```
export interface Article {
```

```
  title: string;
  body: string;
}
```

リスト 3.2.5.2 server/models/app-error.ts

```
export class AppError extends Error {
  status: number;
}
```

3.2.6 エラーページ用テンプレートの作成

エラーページ用のテンプレートを作成します。今回はテンプレートエンジンにEJSを使用します。サンプルアプリのエラーページなので、単純にステータスコードとエラーメッセージを表示するだけの簡素なものを作ります。

リスト 3.2.6 server/views/error.ejs

```
<!doctype html>
<html lang="ja">
<head>
  <meta charset="UTF-8">
  <meta name="viewport" content="width=device-width,
user-scalable=no, initial-scale=1.0, maximum-scale=1.0,
minimum-scale=1.0">
  <meta http-equiv="X-UA-Compatible" content="ie=edge">
  <title>ERROR | <%-status%></title>
</head>
<body>
  <div>
    <%-status%> <%-message%>
  </div>
</body>
</html>
```

3.2.7 Dao クラス

次はDaoクラスを作成します。今回使用するテーブルは一つだけなので、1ファイルで済ませてしまいます。基底クラスやインタフェースも作成しません。

第3章 シングルページアプリケーションの作成 | 31

リスト3.2.7 server/dao/article-dao.ts

```typescript
import * as mysql from 'mysql';
import { Article } from '../models/article';
import { AppError } from '../models/app-error';

/**
 * 記事データ用Daoクラス
 */
export class ArticleDao {

  /**
   * データベースコネクション
   */
  private connection: mysql.IConnection;

  constructor(connection: mysql.IConnection) {
    this.connection = connection;
  }

  insertArticle(article: Article): Promise<number> {
    const query = 'insert into article (title, body, created_at,
updated_at) values (?, ?, now(), now())';
    const param = [
      article.title,
      article.body
    ];
    return new Promise<number>((resolve, reject) => {
      this.connection.query(query, param, (error, result) => {
        if (error) {
          reject(error);
          return;
        }
        resolve(result.insertId);
      });
    });
  }

  selectAllArticles(): Promise<Article[]> {
    return new Promise<Article[]>((resolve, reject) => {
      const query = 'select' +
        ' id' +
        ' ,title' +
```

32　第3章 シングルページアプリケーションの作成

```typescript
          ' ,body' +
          ' ,DATE_FORMAT(created_at, \'%Y-%m-%d %k:%i:%s\') as
createdAt' +
          ' ,DATE_FORMAT(updated_at, \'%Y-%m-%d %k:%i:%s\') as
updatedAt' +
          ' from article' +
          ' order by id';
        this.connection.query(query, [], (error, results) => {
          if (error) {
            reject(error);
            return;
          }
          resolve(results);
        });
      });
    }

    selectCount(): Promise<number> {
      return new Promise<number>((resolve, reject) => {
        const query = 'select count(id) as count from article';
        this.connection.query(query, [], (error, results) => {
          if (error) {
            reject(error);
            return;
          }
          if (typeof results[0].count !== 'number') {
            throw new TypeError();
          }
          resolve(results[0].count);
        });
      });
    }

    selectArticles(offset: number = 0, limit: number = 0):
Promise<Article[]> {
      return new Promise<Article[]>((resolve, reject) => {
        const query = 'select' +
            ' id' +
            ' ,title' +
            ' ,body' +
            ' ,DATE_FORMAT(created_at, \'%Y-%m-%d %k:%i:%s\') as
createdAt' +
```

第3章 シングルページアプリケーションの作成　33

```typescript
            ' ,DATE_FORMAT(updated_at, \'%Y-%m-%d %k:%i:%s\') as
updatedAt' +
        ' from article' +
        ' order by id' +
        ' limit ?, ?';
      this.connection.query(query, [offset, limit], (error, results)
=> {
        if (error) {
          reject(error);
          return;
        }
        resolve(results);
      });
    });
  }

  selectArticleById(id: number): Promise<Article[]> {
    return new Promise<Article[]>((resolve, reject) => {
      const query = 'select ' +
          ' id' +
          ' ,title' +
          ' ,body' +
          ' ,DATE_FORMAT(created_at, \'%Y-%m-%d %k:%i:%s\') as
createdAt' +
          ' ,DATE_FORMAT(updated_at, \'%Y-%m-%d %k:%i:%s\') as
updatedAt' +
        ' from article' +
        ' where id = ?';
      this.connection.query(query, [id], (error, results) => {
        if (error) {
          reject(error);
          return;
        }
        if (!Array.isArray(results) || results.length === 0) {
          const appError = new AppError('Article data is not
found.');
          appError.status = 404;
          reject(appError);
          return;
        }
        resolve(results);
      });
```

```
    });
  }

  updateArticle(id: number, article: Article): Promise<any> {
    return new Promise<any>((resolve, reject) => {
      const query = 'update article ' +
        'set title = ? ' +
          ', body = ? ' +
          ', updated_at = now() ' +
        'where id = ?';
      const params = [
        article.title,
        article.body,
        id
      ];
      this.connection.query(query, params, (error, results) => {
        if (error) {
          reject(error);
          return;
        }
        resolve(results);
      });
    });
  }

  deleteArticle(id: number): Promise<any> {
    return new Promise<void>((resolve, reject) => {
      const query = 'DELETE FROM article WHERE ID = ?';
      this.connection.query(query, [id], (error, result) => {
        if (error) {
          reject(error);
          return;
        }
        resolve(result);
      });
    });
  }

  lock(id: number): Promise<Article> {
    return new Promise<Article>((resolve, reject) => {
      const query = 'select ' +
        ' id ' +
```

```
             ' ,title' +
             ' ,body' +
             ' ,DATE_FORMAT(created_at, \'%Y-%m-%d %k:%i:%s\') as
createdAt' +
             ' ,DATE_FORMAT(updated_at, \'%Y-%m-%d %k:%i:%s\') as
updatedAt' +
             ' from article where id = ? for update';
        this.connection.query(query, [id], (error, results) => {
          if (error) {
            reject(error);
            return;
          }
          resolve(results);
        });
      });
    }

}
```

3.2.8 サービスクラス

次にサービスクラスを作成します。これもDaoクラス同様1クラスのみ作成します。

リスト 3.2.8 server/dao/article-service.ts

```
import * as mysql from 'mysql';
import { DbConfigManager } from '../config/db-config-manager';
import { ArticleDao } from '../dao/article-dao';
import { Article } from '../models/Article';
import { AppError } from '../models/app-error';


export class ArticleService {

  /**
   * 登録済みの全ての記事データを取得します。
   * @returns {Promise<Article[]>}
   */
  async findAllArticles(): Promise<Article[]> {
    const connection =
mysql.createConnection(DbConfigManager.getConfig());
    connection.connect();
```

```
    const articleDao = new ArticleDao(connection);
    try {
      return await articleDao.selectAllArticles();
    } catch(error) {
      throw error;
    } finally {
      connection.destroy();
    }
  }

  /**
   * 第一引数と第二引数で指定された登録済記事データを取得します。
   * @param {number} offset
   * @param {number} limit
   * @returns {Promise<Article[]>}
   */
  async findArticles(offset: number, limit: number):
Promise<Article[]> {
    const connection =
mysql.createConnection(DbConfigManager.getConfig());
    connection.connect();
    const articleDao = new ArticleDao(connection);
    try {
      return await articleDao.selectArticles(offset, limit);
    } catch(error) {
      throw error;
    } finally {
      connection.destroy();
    }
  }

  /**
   * 登録されている記事の件数を取得します。
   * @returns {Promise<number>}
   */
  async findArticleCount(): Promise<number> {
    const connection =
mysql.createConnection(DbConfigManager.getConfig());
    connection.connect();
    const articleDao = new ArticleDao(connection);
    try {
      return await articleDao.selectCount();
```

第3章 シングルページアプリケーションの作成　37

```typescript
    } catch(error) {
      throw error;
    } finally {
      connection.destroy();
    }
  }

  /**
   * 記事を新規登録します。登録完了時、登録した記事のidを返します。
   * @param article
   * @returns {Promise<number>}
   */
  async createArticle(article: Article): Promise<number> {
    const connection =
mysql.createConnection(DbConfigManager.getConfig());
    connection.connect();
    const articleDao = new ArticleDao(connection);
    try {
      return await articleDao.insertArticle(article);
    } catch(error) {
      throw error;
    } finally {
      connection.destroy();
    }
  }

  /**
   * 引数で指定されたidの記事を取得します。
   * @param id {number}
   * @returns {Promise<Article>}
   */
  async findArticle(id: number): Promise<Article[]> {
    const connection =
mysql.createConnection(DbConfigManager.getConfig());
    connection.connect();
    const articleDao = new ArticleDao(connection);
    try {
      return await articleDao.selectArticleById(id);
    } catch(error) {
      throw error;
    } finally {
      connection.destroy();
```

```
    }
  }

  async modifyArticle(id: number, article: Article): Promise<Article>
{
    const connection =
mysql.createConnection(DbConfigManager.getConfig());
    connection.connect();
    const articleDao = new ArticleDao(connection);
    try {
      const results = await articleDao.lock(id);
      if (!Array.isArray(results) || results.length === 0) {
        const error = new AppError();
        error.status = 404;
        throw error;
      }
      if (results[0].updatedAt !== article.updatedAt) {
        const error = new AppError();
        error.status = 500;
        throw error;
      }
      return await articleDao.updateArticle(id, article);
    } catch(error) {
      throw error;
    } finally {
      connection.destroy();
    }
  }

  async removeArticle(id: number): Promise<void> {
    const connection =
mysql.createConnection(DbConfigManager.getConfig());
    connection.connect();
    const articleDao = new ArticleDao(connction);
    try {
      const results = await articleDao.lock(id);
      if (!Array.isArray(results) || results.length === 0) {
        const error = new AppError();
        error.status = 404;
        throw error;
      }
      return await articleDao.deleteArticle(id);
```

```
  } catch(error) {
    throw error;
  } finally {
    connection.destroy();
  }
}
}
```

3.2.9 コントローラクラス

最後にコントローラクラスを作成します。

リスト 3.2.9 server/controllers/article-controller.ts

```
import * as express from 'express';
import { ArticleService } from '../services/ArticleService';
import { AppError } from "../models/AppError";
import { Article } from "../models/Article";
import { RegisteredArticle } from "../models/RegisteredArticle";

/**
 * 記事API用コントローラ
 */
export class ArticleController {

  private articleService: ArticleService;

  /**
   * コンストラクタ
   */
  constructor() {
    this.articleService = new ArticleService();
  }

  all(req: express.Request, res: express.Response, next:
express.NextFunction): void {
    this.articleService.findAllArticles()
      .then((articles) => {
        res.send({data: articles});
      })
      .catch((error) => {
        error.status = error.status || 500;
```

40 | 第3章 シングルページアプリケーションの作成

```typescript
      next(error);
    });
  }

  index(req: express.Request, res: express.Response, next:
express.NextFunction): void {
    const offset = Number.parseInt(req.query.offset, 10);
    const limit = Number.parseInt(req.query.limit, 10);
    if (Number.isNaN(offset) || Number.isNaN(limit)) {
      const error = new AppError('不正なパラメータです。');
      error.status = 400;
      next(error);
      return;
    }
    this.articleService.findArticles(offset, limit)
      .then((articles) => {
        res.send({ data: articles })
      })
      .catch((error) => {
        error.status = error.status || 500;
        next(error);
      });
  }

  count(req: express.Request, res: express.Response, next:
express.NextFunction): void {
    this.articleService.findArticleCount()
      .then((count) => {
        res.send({ data: count });
      })
      .catch((error) => {
        error.status = error.status || 500;
        next(error);
      });
  }

  create(req: express.Request, res: express.Response, next:
express.NextFunction): void {
    const article: Article = req.body;
    if (!article.title) {
      const error = new AppError('タイトルが空です。');
      error.status = 400;
```

```
        next(error);
        return;
      }
      this.articleService.createArticle(article)
        .then((insertId) => {
          res.send({ data: insertId });
        })
        .catch((error) => {
          error.status = error.status || 500;
          next(error);
        });
    }

    read(req: express.Request, res: express.Response, next:
  express.NextFunction): void {
      const id = Number.parseInt(req.params.id, 10);
      if (Number.isNaN(id)) {
        const error = new AppError();
        error.status = 400;
        next(error);
        return;
      }
      this.articleService.findArticle(id)
        .then((result) => {
          res.send({ data: result });
        })
        .catch((error) => {
          error.status = error.status || 500;
          next(error);
        });
    }

    update(req: express.Request, res: express.Response, next:
  express.NextFunction): void {
      const id = Number.parseInt(req.params.id, 10);
      const article: RegisteredArticle = req.body;
      if (!article.title) {
        const error = new AppError('タイトルが空です。');
        error.status = 400;
        next(error);
        return;
      }
```

```typescript
    if (Number.isNaN(id)) {
      const error = new AppError();
      error.status = 400;
      next(error);
      return;
    }
    this.articleService.modifyArticle(id, article)
      .then((result) => {
        res.send({ data: result });
      })
      .catch((error) => {
        error.status = error.status || 500;
        next(error);
      });
  }

  delete(req: express.Request, res: express.Response, next:
express.NextFunction): void {
    const id = Number.parseInt(req.params.id, 10);
    if (Number.isNaN(id)) {
      const error = new AppError();
      error.status = 400;
      next(error);
      return;
    }
    this.articleService.removeArticle(id)
      .then((result) => {
        res.send({ data: result });
      })
      .catch((error) => {
        error.status = error.status || 500;
        next(error);
      });
  }

}
```

3.3 フロントエンドの開発

サーバーサイドの開発が一通り完了したので、次はフロントエンドの開発を進めていきます。

3.3.1 Angular-CLIのインストール

Angularのバージョン2以降では、Angular CLIというツールを使用することができます。Angular CLIは、Angularを使用したアプリケーションの雛形や、コンポーネントの作成などを補助するコマンドラインツールです。

Angular CLIを使用するには、2章でTypeScriptをインストール時と同様に、npmを使用して、グローバルにインストールします。

リスト3.3.1.1

```
$ npm install angular-cli --global
```

なお、本書ではAngular CLIのバージョン1.6.3を使用して検証しています。明示的にバージョンを指定してインストールする場合、リスト3.3.1.2のようにバージョンを指定します。

リスト3.3.1.2

```
$ npm install angular-cli@1.6.3 --global
```

正常にインストールしたかどうか確認するために、リスト3.3.1.3のコマンドを実行し、angular-cliのバージョンを確認しましょう。

リスト3.3.1.3

```
$ ng --version
```

3.3.2 プロジェクトの作成

Angular-CLIのインストールが完了したら、chapter-3ディレクトリ内にclientという名前でAngularのプロジェクトを作成します。リスト3.3.2のコマンドを実行しましょう。

リスト3.3.2

```
$ cd [ワーキングディレクトリ『chapter-3』のパス]
$ ng new client
```

完了したら、chapter-3ディレクトリにclientという名前でプロジェクトディレクトリが作成されているはずです。

44 | 第3章 シングルページアプリケーションの作成

3.3.3　その他のライブラリのインストール

次に、Angular CLIで作成された雛形ではインストールされていないライブラリをインストールしていきます。今回は、見た目を多少整えるために以下の3つをインストールします。

- Font Awesome
    - —UI上に表示するアイコンをWebフォントとして使用できるようにしたものです。
- Pure.css
    - —CSSフレームワークです。BootstrapやFoundationと異なり、CSSファイルのみで完結しています。
- toastr
    - —簡単にトースト通知を実装できるライブラリです。

リスト3.3.3

```
$ cd [ワーキングディレクトリ『chapter-3』のパス]/client
$ npm install font-awesome --save
$ npm install pure-css --save
$ npm install toastr --save
$ npm install @types/toastr --save-dev
```

3.3.4 .angular-cli.jsonの修正

次にclientディレクトリ直下の.angular-cli.jsonの内容を修正します。ここでは、ビルド後のファイルを出力するディレクトリを変更します。デフォルトでは"dist"が指定されているapps.outDirの値を、"../server/public"に修正します。

リスト3.3.4

```
{

  // 省略

  "apps": [
    {
      "root": "src",
      "outDir": "../server/public",
```

3.3.5 プロキシの設定

clientディレクトリ直下にproxy.conf.jsonという名前で、リストの内容のファイルを作

第3章 シングルページアプリケーションの作成　45

成します。

リスト3.3.5

```
{
  "/api": {
    "target": "http://localhost:3000",
    "secure": false
  }
}
```

また、`package.json`の scripts項目の start を以下のように修正します。

リスト3.3.2

```
"start": "ng serve --proxy-config proxy.conf.json",
```

　Angularの開発においては、開発中はhttp://localhost:4200にアクセスしながら作業します。しかし、実際に作成したAPIサーバーのオリジンはhttp://localhost:3000のため、/apiにリクエストがきた場合、 http://localhost:3000にフォワードするための設定です。

3.3.6　開発用タスクの実行

　ここまで設定ができたら、APIサーバーと、Angularの開発用スクリプトを実行し開発を始めましょう。

　まずはAPIサーバーの開発用スクリプトを実行します。

リスト3.3.6.1

```
$ cd [ワーキングディレクトリ『chapter-3』のパス]/server
$ npm run dev
```

別なターミナルを開き以下のコマンドを実行し、Angular側の開発用スクリプトを実行します。

リスト3.3.6.2

```
$ cd [ワーキングディレクトリ『chapter-3』のパス]/client
$ npm run start
```

それぞれ、正常に実行できたら以下のURLにブラウザからアクセスします。

```
http://localhost:4200
```

以下の画面が表示されたら、ここまでの設定は完了です。

3.3.7 app-moduleの修正

client直下のapp.module.tsを以下のように修正します。NgModuleデコレータのdeclarationsにコンポーネントを、importsに外部のNgModuleを、providersにDI対象のモジュールを指定している点です。今の時点では、まだこれらのファイルを作成していませんが、全てのファイルを作成し終えたら、それぞれどの内容のファイルを指定しているのか確認すると良いでしょう。

リスト3.3.7

```
import { BrowserModule } from '@angular/platform-browser';
import { NgModule } from '@angular/core';
import { RouterModule } from '@angular/router';
import { HttpClientModule } from '@angular/common/http';

import { AppStore } from './app-store';

import { RootComponent } from './components/root.component';
import { ArticleDetailPageComponent } from
```

```
'./components/article-detail-page/article-detail-page.component';
import { ArticleListPageComponent } from
'./components/article-list-page/article-list-page.component';
import { RegisterArticlePageComponent } from
'./components/register-article-page/register-article-page.component';
import { UpdateArticlePageComponent } from
'./components/update-article-page/update-article-page.component';

import { ArticleListPageService } from
'./services/article-list-page.service';
import { ArticleDetailPageService } from
'./services/article-detail-page.service';
import { UpdateArticlePageService } from
'./services/update-article-page.service';
import { RegisterArticlePageService } from
'./services/register-article-page.service';

import { ArticleDetailPageResolver } from
'./resolvers/article-detail-page.resolver';
import { ArticleListPageResolver } from
'./resolvers/article-list-page.resolver';
import { UpdateArticlePageResolver } from
'./resolvers/update-article-page.resolver';
import { ArticleCardComponent } from
'./components/article-list-page/article-card/article-card.component';

@NgModule({
  declarations: [
    RootComponent,
    ArticleDetailPageComponent,
    ArticleListPageComponent,
    RegisterArticlePageComponent,
    UpdateArticlePageComponent,
    ArticleCardComponent
  ],
  imports: [
    BrowserModule,
    HttpClientModule,
    RouterModule.forRoot([
      {
        path: '',
        component: ArticleListPageComponent,
```

```
      resolve: {
        articleList: ArticleListPageResolver
      }
    },
    {
      path: 'create-article',
      component: RegisterArticlePageComponent
    },
    {
      path: 'article-detail/:id',
      component: ArticleDetailPageComponent,
      resolve: {
        article: ArticleDetailPageResolver
      }
    },
    {
      path: 'update-article/:id',
      component: UpdateArticlePageComponent,
      resolve: {
        article: UpdateArticlePageResolver
      }
    }
  ])
  ],
  providers: [
    AppStore,
    ArticleListPageService,
    ArticleDetailPageService,
    UpdateArticlePageService,
    RegisterArticlePageService,
    ArticleDetailPageResolver,
    ArticleListPageResolver,
    UpdateArticlePageResolver
  ],
  bootstrap: [RootComponent]
})
export class AppModule { }
```

3.3.8 モデルの作成

サーバーサイドの開発時と同様にモデルを定義します。それぞれ、記事、フロントエンドの
アプリケーションの状態、APIのレスポンスの型を作成しています。

リスト 3.3.8.1 client/app/src/model/article.ts

```
export interface Article {
  id?: number,
  title: string;
  body: string;
  createdAt?: string,
  updatedAt?: string
}
```

リスト 3.3.8.2 client/app/src/models/app-state.ts

```
import { Article } from './article';
export interface AppState {
  articles?: Article[];
  showLoading?: boolean;
  currentShowArticle?: Article;
  currentUpdateArticle?: Article;
}
```

リスト 3.3.8.3 client/app/src/models/response-body.ts

```
export interface ResponseBody<T> {
  data?: T;
  message?: string;
}
```

3.3.9 グローバルCSSの設定

画面を確認できるようになったら、まずはグローバルに適用されるCSSの設定をしましょう。
インストールしたPure CSSやFont Awesomeは、ここで読み込むよう修正します。Client直下
のstyles.cssを以下のように修正します。

リスト 3.3.9.1 client/app/src/style.css

```
/* You can add global styles to this file, and also import other
style files */

/* purecss */
```

50 │ 第3章 シングルページアプリケーションの作成

```
@import '../node_modules/purecss/build/pure-min.css';
@import '../node_modules/purecss/build/grids-responsive-min.css';

/* font-awesome */
@import '../node_modules/font-awesome/css/font-awesome.min.css';

/* toastr */
@import '../node_modules/toastr/build/toastr.min.css';
```

さらに、ボタンの色や、通信中のスピナーの見た目などを調整するための設定も追記します。

リスト 3.3.9.2 client/app/src/style.css

```
body {
  background-color: #ececec;
}

.spinner-background {
  position: fixed;
  top: 0;
  left: 0;
  height: 100vh;
  width: 100vw;
}

.spinner-container {
  width: 200px;
  height: 100px;
  border-radius: 10px;
  background: #fff;
  padding-top: 20px;
  position: fixed;
  top: calc(50vh - 50px);
  text-align: center;
  color: #888;
  left: calc(50vw - 100px);
}

/* ボタン */
.button-success,
.button-error,
.button-warning,
.button-secondary {
```

```css
  color: white;
  border-radius: 4px;
  text-shadow: 0 1px 1px rgba(0, 0, 0, 0.2);
}

.button-success {
  background: rgb(28, 184, 65); /* this is a green */
}

.button-error {
  background: rgb(202, 60, 60); /* this is a maroon */
}

.button-warning {
  background: rgb(223, 117, 20); /* this is an orange */
}

.button-secondary {
  background: rgb(66, 184, 221); /* this is a light blue */
}

.button-xsmall {
  font-size: 70%;
}

.button-small {
  font-size: 85%;
}

.button-large {
  font-size: 110%;
}

.button-xlarge {
  font-size: 125%;
}
```

3.3.10 index.htmlの修正

client直下のindex.htmlを以下のように修正します。このindex.htmlがシングルページ
アプリケーションのエントリポイントとなります。ポイントは2行目のhtmlタグのlang属性

52 | 第3章 シングルページアプリケーションの作成

を変更している点と、app-rootタグ内に読み込み中に表示するスピナー要素を追加している点です。

リスト3.2.10 client/index.html

```html
<!doctype html>
<html lang="ja">
<head>
  <meta charset="utf-8">
  <title>Sample CRUD Application</title>
  <base href="/">
  <meta name="viewport" content="width=device-width,
initial-scale=1">
  <link rel="icon" type="image/x-icon" href="favicon.ico">
</head>
<body>
  <app-root>
    <div class="spinner-background"></div>
    <div class="spinner-container">
      <i class="fa fa-spinner fa-spin fa-2x fa-fw"></i>
      <p>Loading...</p>
    </div>
  </app-root>
</body>
</html>
```

3.3.11 コンポーネントの追加

次にコンポーネントクラスを作成します。コンポーネントとは画面を構成する部品のことを指します。VelocityやSmartyといったテンプレートと異なる点として、テンプレートはあくまで静的な見た目を定義するものですが、コンポーネントは見た目に加えボタンを押した時の処理などの振る舞いも定義します。

Angularでのコンポーネントは、TypeScript、HTML、CSSの3つから構成されます。これらは1つのファイルにすべて記述することもできますし、それぞれ別のファイルにすることもできます。今回はすべて別のファイルに記述していきます。HTML、CSSへの参照方法はTypeScriptファイルの『@Component』デコレータ内の『templateUrl』、『styleUrls』属性で指定します。

今回は以下の6つのコンポーネントを作成します。

・root.component
　―アプリケーション全体を包括するコンポーネント
・article-list-page.component

―記事一覧表示部分のコンポーネント

・article-card.component

　―article-list-page.componentの子コンポーネント。カードレイアウトになっている各記事の
　要素です。

・register-article-page.component

　―記事作成フォームのコンポーネント

・update-article-page.component

　―記事更新フォームのコンポーネント

・article-detail-page.component

　―記事内容表示領域のコンポーネント

3.3.11.1 root.component

リスト 3.3.11.1.1 client/app/src/components/root.component.ts

```
import { Component, OnInit, OnDestroy } from '@angular/core';
import { AppStore } from '../app-store';
import { AppState } from '../models/app-state';

@Component({
  selector: 'app-root',
  templateUrl: './root.component.html',
  styleUrls: ['./root.component.css']
})
export class RootComponent implements OnInit, OnDestroy {

  showLoading: boolean;

  constructor(private appStore: AppStore) {
    this.showLoading = this.appStore.appState.showLoading;
    this.changeLoader = this.changeLoader.bind(this);
  }

  ngOnInit() {
    this.appStore.registerHandler('CHANGE.LOADER',
this.changeLoader);
  }

  ngOnDestroy() {
    this.appStore.removeHandler('CHANGE.LOADER', this.changeLoader);
  }
```

54 ｜ 第3章 シングルページアプリケーションの作成

```
  changeLoader(eventName: string, beforeAppState: AppState,
currentAppeState: AppState) {
    this.showLoading = currentAppeState.showLoading;
  }
}
```

リスト 3.3.11.1.2 client/app/src/components/root.component.html

```
<header class="header">
  <h1 class="title"><a class="link" href="/">Sample App</a></h1>
  <a class="pure-button pure-button-primary button-xsmall"
[routerLink]="['/create-article']">
    <i class="fa fa-file-o" aria-hidden="true"></i> 新規作成
  </a>
</header>

<div class="content">
  <router-outlet></router-outlet>
</div>

<div class="spinner-background" *ngIf="showLoading"></div>
<div class="spinner-container" *ngIf="showLoading">
  <i class="fa fa-spinner fa-spin fa-2x fa-fw"></i>
  <p>Loading...</p>
</div>
```

リスト 3.3.11.1.2 client/app/src/components/root.component.html

```
.header {
  background-color: #fff;
  box-shadow: 0 0 0 0 #ccc, 0 5px 5px 0 #ccc;
  padding: 10px;
  position: fixed;
  width: 100vw;
}

.title {
  font-size: 16px;
  margin: 0 0 5px 0;
}

.link {
  color: #888;
  text-decoration: none;
```

第3章 シングルページアプリケーションの作成 | 55

```
}

.content {
  padding: 90px 10px 0 10px;
}
```

3.3.11.2 article-list-page.component

リスト 3.3.11.2.1 client/app/src/components/article-list-page /article-list-page.component.ts

```
import { Component, OnInit, OnDestroy } from '@angular/core';
import { ActivatedRoute } from '@angular/router';

import { AppStore } from '../../app-store';

import { AppState } from '../../models/app-state';
import { Article } from '../../models/article';

@Component({
  selector: 'app-article-list-page',
  templateUrl: './article-list-page.component.html',
  styleUrls: ['./article-list-page.component.css']
})
export class ArticleListPageComponent implements OnInit, OnDestroy {

  articles: Article[];

  constructor(
    private appStore: AppStore,
    private route: ActivatedRoute
  ) {
    this.articles = this.route.snapshot.data.articleList.data;
    this.onChangeArticles = this.onChangeArticles.bind(this);
  }

  ngOnInit() {
    this.appStore.registerHandler('CHANGE.ARTICLES',
this.onChangeArticles);
  }

  ngOnDestroy() {
    this.appStore.removeHandler('CHANGE.ARTICLES',
```

```
    this.onChangeArticles);
  }

  onChangeArticles(eventName, beforeAppState, appState: AppState) {
    this.articles = appState.articles;
  }
}
```

リスト 3.3.11.2.2 client/app/src/components/article-list-page /article-list-page.component.html

```html
<div *ngIf="articles.length === 0">登録された記事がありません。</div>

<div class="pure-g">
  <div class="pure-u-1 pure-u-md-1-4" *ngFor="let article of
articles">
    <app-article-row [article]="article"></app-article-row>
  </div>
</div>
```

リスト 3.3.11.2.2 client/app/src/components/article-list-page /article-list-page.component.css

```css
.articles {
  width: 100%;
}
```

3.3.11.3 article-card.component

リスト 3.3.11.3.1 client/app/src/components/article-list-page/ article-card/article-card.component.ts

```typescript
import { Component, OnInit, Input } from '@angular/core';
import * as moment from 'moment';
import * as toastr from 'toastr';
import { Article } from '../../../models/article';
import { ArticleListPageService } from
'../../../services/article-list-page.service';

@Component({
  selector: 'app-article-row',
  templateUrl: './article-card.component.html',
  styleUrls: ['./article-card.component.css']
})
export class ArticleCardComponent implements OnInit {
```

第3章 シングルページアプリケーションの作成 | 57

```
@Input() article: Article;

title: string;

formattedCreatedAt: string;

formattedUpdatedAt: string;

constructor(
  private service: ArticleListPageService
) { }

ngOnInit() {
  const INPUT_FORMAT = 'YYYY-MM-DD HH:mm:ss';
  const OUTPUT_FORMAT = 'YYYY年MM月DD日';
  this.formattedCreatedAt = moment(this.article.createdAt,
INPUT_FORMAT).format(OUTPUT_FORMAT);
  this.formattedUpdatedAt = moment(this.article.updatedAt,
INPUT_FORMAT).format(OUTPUT_FORMAT);
}

deleteButtonClicked(articleId: number) {
  if ( ! confirm('削除しますか?')) {
    return;
  }
  this.service.deleteArticle(articleId)
    .subscribe(() => {
      toastr.success('記事を削除しました。');
    },
    (error) => {
      const message = error.message || 'エラーが発生しました。';
      toastr.error(message);
    });
}
}
```

リスト 3.3.11.3.1 client/app/src/components/article-list-page/ article-card/article-card.component.html

```
<div class="panel">

  <section class="date-section">
    <span class="article-created-at">作成日：{{ formattedCreatedAt
```

```html
}}</span><br>
    <span class="artilce-updated-at">更新日：{{ formattedUpdatedAt
}}</span>
  </section>

  <h2 class="article-title">
    <a class="article-title-link" [routerLink]="['/article-detail',
article.id]">{{ article.title }}</a>
  </h2>

  <section class="article-body">本文：{{ article.body }}</section>

  <section>
    <a class="pure-button button-secondary"
[routerLink]="['/update-article', article.id]">
      <i class="fa fa-pencil-square-o" aria-hidden="true"></i>
    </a>
    <button class="pure-button button-error"
(click)="deleteButtonClicked(article.id)">
      <i class="fa fa-trash-o" aria-hidden="true"></i>
    </button>
  </section>
</div>
```

リスト 3.3.11.3.1 client/app/src/components/article-list-page/ article-card/article-card.component.css

```css
panel {
  background-color: #fff;
  box-shadow:0 0 0 0 #caccca,0 3px 5px 0 #919191;
  border: 1px solid #cccccc;
  box-sizing: border-box;
  margin: 5px;
  padding: 10px;
}

.date-section {
  font-size: 12px;
  text-align: right;
}
```

```css
.article-title {
  font-size: 16px;
  overflow: hidden;
  margin: 10px 0 0 0;
  text-overflow: ellipsis;
  width: 100%;
  white-space: nowrap;
}

.article-title-link {
  color: #888;
  text-decoration: none;
}

.article-body {
  font-size: 12px;
  margin: 5px 0 5px 0;
  overflow: hidden;
  text-overflow: ellipsis;
  width: 100%;
  white-space: nowrap;
}
```

3.3.11.3 register-article.component

リスト 3.3.11.3.1 client/app/src/components/register-article /register-article.component.ts

```typescript
import { Component, OnInit, OnDestroy } from '@angular/core';
}
```

リスト 3.3.11.3.2 client/app/src/components/register-article /register-article.component.html

```typescript
import { Component, OnInit } from '@angular/core';

import * as toastr from 'toastr';

import { Article } from '../../models/article';

import { RegisterArticlePageService } from
'../../services/register-article-page.service';
```

```typescript
@Component({
  selector: 'app-register-article-page',
  templateUrl: './register-article-page.component.html',
  styleUrls: ['./register-article-page.component.css']
})
export class RegisterArticlePageComponent implements OnInit {

  article: Article;

  isCompleted: boolean;

  constructor(private service: RegisterArticlePageService) { }

  ngOnInit() {
    this.article = {
      title: '',
      body: ''
    };
    this.isCompleted = false;
  }

  titleChanged(event: Event) {
    this.article.title = (event.target as HTMLInputElement).value;
  }

  bodyChanged(event: Event) {
    this.article.body = (event.target as HTMLInputElement).value;
  }

  registerButtonClicked() {
    this.service.createArticle(this.article)
      .subscribe(() => {
        this.isCompleted = true;
        toastr.success('記事を登録しました。');
      },
      (error) => {
        const message = error.message || 'エラーが発生しました。';
        toastr.error(message);
      });
  }
}
```

第3章 シングルページアプリケーションの作成 | 61

リスト 3.3.11.4.3 client/app/src/components/register-article /register-article.component.css

```
<a [routerLink]="['/']">もどる</a>

<div class="complete-mesage" *ngIf="isCompleted">
  <div>記事の登録が完了しました。</div>
  <p><a [routerLink]="['/']">一覧にもどる</a></p>
</div>

<div class="pure-form pure-form-aligned" *ngIf="!isCompleted">
  <h2>記事の作成</h2>
  <div class="pure-control-group">
    <label>タイトル</label>
    <input type="text" class="article__title" placeholder="タイトル"
        value="{{ article.title }}" (change)="titleChanged($event)">
  </div>
  <div class="pure-control-group">
    <label>本文</label>
    <textarea class="article__body" placeholder="本文"
(change)="bodyChanged($event)">{{ article.body }}</textarea>
  </div>
  <div class="pure-controls">
    <button class="pure-button pure-button-primary"
(click)="registerButtonClicked()">登録</button>
  </div>
</div>
```

リスト 3.3.11.4.3 client/app/src/components/register-article /register-article.component.css

```
.complete-mesage {
  background: #fff;
  border-radius: 10px;
  margin: 10px;
  text-align: center;
  padding: 20px;
}
```

3.3.11.5 update-article-page.component

/update-article-page.component.ts

```typescript
import { Component } from '@angular/core';
import { ActivatedRoute } from '@angular/router';

import * as toastr from 'toastr';

import { Article } from '../../models/article';

import { UpdateArticlePageService } from
'../../services/update-article-page.service';

@Component({
  selector: 'app-update-article-page',
  templateUrl: './update-article-page.component.html',
  styleUrls: ['./update-article-page.component.css']
})
export class UpdateArticlePageComponent {

  isCompleted: boolean;

  article: Article;

  constructor(
    private route: ActivatedRoute,
    private service: UpdateArticlePageService
  ) {
    this.article = this.route.snapshot.data.article.data[0];
  }

  titleChanged(event) {
    this.article.title = event.target.value;
  }

  bodyChanged(event) {
    this.article.body = event.target.value;
  }

  updateButtonClicked() {
    this.service.updateArticle(this.article)
      .subscribe(() => {
```

```
      this.isCompleted = true;
      toastr.success('記事を更新しました。');
    },
    (error) => {
      const message = error.message || 'エラーが発生しました。';
      toastr.error(message);
    });
  }
}
```

リスト3.3.11.5.2 client/app/src/components/update-article-page /update-article-page.component.html

```
<a [routerLink]="['/']">もどる</a>

<div class="complete-mesage" *ngIf="isCompleted">
  <div>記事の更新が完了しました。</div>
  <p><a [routerLink]="['/']">一覧にもどる</a></p>
</div>

<div class="pure-form pure-form-aligned" *ngIf="!isCompleted">
  <h2>記事の更新</h2>
  <div class="pure-control-group">
    <label>タイトル</label>
    <input type="text" class="article__title" placeholder="タイトル"
           value="{{ article.title }}"
(change)="titleChanged($event)">
  </div>
  <div class="pure-control-group">
    <label>本文</label>
    <textarea class="article__body" placeholder="本文"
(change)="bodyChanged($event)">{{ article.body }}</textarea>
  </div>
  <div class="pure-controls">
    <button class="pure-button pure-button-primary"
(click)="updateButtonClicked()">登録</button>
  </div>
</div>
```

リスト3.3.11.5.3 client/app/src/components/ update-article-page /update-article-page.component.css

```
.complete-mesage {
  background: #fff;
```

64 | 第3章 シングルページアプリケーションの作成

```css
  border-radius: 10px;
  margin: 10px;
  text-align: center;
  padding: 20px;
}
```

3.3.11.6 article-detail.component

リスト 3.3.11.6.1 client/app/src/components/article-detail /article-detail.component.ts

```typescript
import { Component } from '@angular/core';
import { ActivatedRoute } from '@angular/router';

import { Article } from '../../models/article';

@Component({
  selector: 'app-article-detail-page',
  templateUrl: './article-detail-page.component.html',
  styleUrls: ['./article-detail-page.component.css']
})
export class ArticleDetailPageComponent {

  article: Article;

  constructor(private route: ActivatedRoute) {
    this.article = this.route.snapshot.data.article.data[0];
  }
}
```

リスト 3.3.11.6.2 client/app/src/components/article-detail /article-detail.component.html

```html
<p>
  <a [routerLink]="['/']">もどる</a>
</p>

<article *ngIf="article" class="">
  <h2>{{ article.title }}</h2>
  <p class="article__body">{{ article.body }}</p>
</article>
```

リスト 3.3.11.6.3 client/app/src/components/article-detail/article-detail.component.css

```css
.article__body {
```

第3章 シングルページアプリケーションの作成 | 65

```
  height: calc(100% - 190px);
  white-space: pre-wrap;
}
```

3.3.12 サービスクラスの作成

コンポーネントを作成したら、次にサービスクラスを作成します。フロントエンドのサービスクラスの役割は、主にサーバーサイドとの通信です。

今回は画面毎に、サービスクラスを作成します。また、各サービスクラスで共通して使用するエラーハンドラ関数モジュールと、リクエストを送信する際のオプションも同じディレクトリ内に作成します。

リスト 3.3.12.1 /client/src/app/src/services/article-list-page.service.ts

```
import { Injectable } from '@angular/core';
import { HttpClient } from '@angular/common/http';

import 'rxjs/add/operator/do';
import 'rxjs/add/operator/mergeMap';
import 'rxjs/add/operator/catch';
import 'rxjs/add/operator/finally';

import { handleError } from './handle-error';
import { requestOptions } from './request-options';
import { AppStore } from '../app-store';

import { Article } from '../models/article';
import { ResponseBody } from '../models/response-body';

@Injectable()
export class ArticleListPageService {

  constructor(
    private httpClient: HttpClient,
    private appStore: AppStore
  ) { }

  findArticles() {
    this.appStore.applyAppState('CHANGE.LOADER', { showLoading: true
});
    return
```

```
this.httpClient.get<ResponseBody<Article[]>>('/api/v1/article/all.json',
requestOptions)
      .do((response) => {
        this.appStore.applyAppState('CHANGE.ARTICLES', { articles:
response.data });
      })
      .catch(handleError)
      .finally(() => {
        this.appStore.applyAppState('CHANGE.LOADER', { showLoading:
false });
      });
  }

  deleteArticle(articleId: number) {
    this.appStore.applyAppState('CHANGE.LOADER', { showLoading: true
});
    const url = '/api/v1/article/${articleId}.json';
    return this.httpClient.delete(encodeURI(url), requestOptions)
      .mergeMap(() =>
this.httpClient.get<ResponseBody<Article[]>>('/api/v1/article/all.json',
requestOptions))
      .do((response) => {
        this.appStore.applyAppState('CHANGE.ARTICLES', { articles:
response.data });
      })
      .catch(handleError)
      .finally(() => {
        this.appStore.applyAppState('CHANGE.LOADER', { showLoading:
false });
      });
  }
}
```

リスト3.3.12.1 client/app/src/services/article-detail-page.service.ts

```
import { Injectable } from '@angular/core';
import { HttpClient } from '@angular/common/http';

import 'rxjs/add/operator/do';
import 'rxjs/add/operator/catch';
import 'rxjs/add/operator/finally';

import { requestOptions } from './request-options';
```

第3章 シングルページアプリケーションの作成 | 67

```
import { handleError } from './handle-error';

import { AppStore } from '../app-store';

import { Article } from '../models/article';
import { ResponseBody } from '../models/response-body';

@Injectable()
export class ArticleDetailPageService {

  constructor(
    private httpClient: HttpClient,
    private appStore: AppStore
  ) { }

  findArticleById(id: number) {
    this.appStore.applyAppState('CHANGE.LOADER', { showLoading: true
});
    const url = '/api/v1/article/${id}.json';
    return this.httpClient.get<ResponseBody<Article>>(encodeURI(url),
requestOptions)
      .do((response) => {
        this.appStore.applyAppState('', { currentShowArticle:
response.data });
      })
      .catch(handleError)
      .finally(() => {
        this.appStore.applyAppState('CHANGE.LOADER', { showLoading:
false });
      });
  }
}
```

リスト 3.3.12.2 client/app/src/services/register-article-page.service.ts

```
import { Injectable } from '@angular/core';
import { HttpClient } from '@angular/common/http';

import 'rxjs/add/operator/catch';
import 'rxjs/add/operator/finally';

import { AppStore } from '../app-store';
```

68 | 第3章 シングルページアプリケーションの作成

```
import { handleError } from './handle-error';
import { requestOptions } from './request-options';

import { Article } from '../models/article';
import { ResponseBody } from '../models/response-body';

@Injectable()
export class RegisterArticlePageService {

  constructor(
    private httpClient: HttpClient,
    private appStore: AppStore
  ) { }

  createArticle(article: Article) {
    this.appStore.applyAppState('CHANGE.LOADER', { showLoading: true
});
    return
this.httpClient.post<ResponseBody<Article>>('/api/v1/article',
article, requestOptions)
      .catch(handleError)
      .finally(() => this.appStore.applyAppState('CHANGE.LOADER', {
showLoading: false }));
  }
}
```

リスト 3.3.12.3 client/app/src/services/update-article-page.service.ts

```
import { Injectable } from '@angular/core';
import { HttpClient } from '@angular/common/http';

import 'rxjs/add/operator/do';
import 'rxjs/add/operator/mergeMap';
import 'rxjs/add/operator/catch';
import 'rxjs/add/operator/finally';

import { handleError } from './handle-error';
import { requestOptions } from './request-options';

import { AppStore } from '../app-store';

import { Article } from '../models/article';
```

第3章 シングルページアプリケーションの作成 | 69

```typescript
import { ResponseBody } from '../models/response-body';

@Injectable()
export class UpdateArticlePageService {

  constructor(
    private httpClient: HttpClient,
    private appStore: AppStore
  ) { }

  findArticleById(id: number) {
    this.appStore.applyAppState('CHANGE.LOADER', { showLoading: true
});
    const url = '/api/v1/article/${id}.json';
    return this.httpClient.get<ResponseBody<Article>>(encodeURI(url),
requestOptions)
      .do((response) => {
        this.appStore.applyAppState('', { currentShowArticle:
response.data });
      })
      .catch(handleError)
      .finally(() => {
        this.appStore.applyAppState('CHANGE.LOADER', { showLoading:
false });
      });
  }

  updateArticle(article: Article) {
    this.appStore.applyAppState('CHANGE.LOADER', { showLoading: true
});
    const url = '/api/v1/article/${article.id}.json';
    return this.httpClient.put<ResponseBody<Article>>(encodeURI(url),
article, requestOptions)
      .do((response) => {
        this.appStore.applyAppState('', { currentShowArticle:
response.data });
      })
      .catch(handleError)
      .finally(() => {
        this.appStore.applyAppState('CHANGE.LOADER', { showLoading:
false });
      });
```

```
    }
}
```

リスト3.3.12.3 client/app/src/services/handle-error.ts

```
export const handleError = function (error: any) {
  const errorMessage = error.message || error.json().message ||
'Server error';
  throw new Error(errorMessage);
};
```

リスト3.3.12.3 client/app/src/services/request-options.ts

```
export const requestOptions = {
  headers: {
    'x-requested-with': 'XMLHttpRequest'
  }
};
```

3.3.13 Storeの作成

　フロント側のアプリケーションの状態を管理するStoreクラスを作成します。このStoreはアプリケーションの状態に変更があった場合に、予めハンドラーとして登録されているコンポーネント側の処理を呼び出す役割も持っています。これにより、常にアプリケーションの状態とコンポーネント側の表示を常に一致させることができます。

リスト3.3.13 client/app/app-store.ts

```
import { Injectable } from '@angular/core';
import { AppState } from './models/app-state';

type Handler = (eventName: string, beforeState: AppState,
currentState: AppState) => void;

interface EmitInfo {

  eventName: string;

  handlers: Handler[];
}

@Injectable()
export class AppStore {
```

第3章 シングルページアプリケーションの作成 | 71

```typescript
appState: AppState;

emitInfoList: EmitInfo[];

constructor() {
  this.appState = {
    articles: [],
    showLoading: false,
  };
  this.emitInfoList = [];
}

applyAppState(eventName: string, chagedState: AppState): void {
  const emitInfo = this.emitInfoList.find((emitInfo: EmitInfo) => {
    return emitInfo.eventName === eventName
  });
  if ( ! emitInfo) {
    return;
  }
  const beforeAppState = Object.assign({}, this.appState);
  this.appState = Object.assign(this.appState, chagedState);
  emitInfo.handlers.forEach((handler: Handler) => {
    handler(eventName, beforeAppState, this.appState);
  });
}

registerHandler(eventName: string, handler: Handler): void {
  const emitInfo = this.emitInfoList.find((emitInfo: EmitInfo) => {
    return emitInfo.eventName === eventName
  });
  if (emitInfo) {
    emitInfo.handlers.push(handler);
  } else {
    this.emitInfoList.push({
      eventName: eventName,
      handlers: [handler]
    });
  }
}

removeHandler(eventName: string, handler: Handler): void {
```

```
    const emitInfo = this.emitInfoList.find((emitInfo: EmitInfo) => {
      return emitInfo.eventName === eventName;
    });
    const handlerIndex = emitInfo.handlers.indexOf(handler);
    if (handlerIndex === -1) {
      return;
    }
    emitInfo.handlers.splice(handlerIndex, 1);
  }
}
```

3.3.14 Resolverクラスの作成

Resolverクラスは、フロントエンドでのルーティングにおいて、遷移先のコンポーネントの初期化の前に、値を取得する処理役割を担います。ここでは、記事一覧画面、記事詳細画面、記事更新画面用にそれぞれクラスを定義します。

リスト3.3.14.1 client/src/app/resolvers/article-detail-page.resolver.ts

```
import { Injectable } from '@angular/core';
import { Resolve, ActivatedRouteSnapshot } from '@angular/router';
import { ArticleDetailPageService } from
'../services/article-detail-page.service';

@Injectable()
export class ArticleDetailPageResolver implements Resolve<any> {

  constructor(private service: ArticleDetailPageService) {
  }

  resolve(route: ActivatedRouteSnapshot) {
    return this.service.findArticleById(route.params.id);
  }
}
```

リスト3.3.14.2 client/src/app/resolvers/article-list-page.resolver.ts

```
import { Injectable } from '@angular/core';
import { Resolve } from '@angular/router';
import { ArticleListPageService } from
'../services/article-list-page.service';
```

```
@Injectable()
export class ArticleListPageResolver implements Resolve<any> {

  constructor(private service: ArticleListPageService) {
  }

  resolve() {
    return this.service.findArticles();
  }
}
```

リスト3.3.14.3 client/src/app/resolvers/article-list-page.resolver.ts

```
import { Injectable } from '@angular/core';
import { Resolve } from '@angular/router';
import { ArticleListPageService } from
'../services/article-list-page.service';

@Injectable()
export class ArticleListPageResolver implements Resolve<any> {

  constructor(private service: ArticleListPageService) {
  }

  resolve() {
    return this.service.findArticles();
  }
}
```

3.4 スクリプト実行手順

　サーバーサイド、フロントエンドそれぞれの`package.json`の`scripts`には、開発時、本番用のスクリプトが設定されています。ここでは開発時と本番用に分けてどのスクリプトを実行するのかを確認します。

3.4.1 開発用スクリプトの実行手順

　3.3.6でもこの手順については触れていますが、再度確認します。開発中は、サーバーサイド、フロントエンドともに、TypeScriptのファイルの変更を監視してリロードするためのタスクを実行します。

　サーバーサイドの開発時はリスト3.4.1.1のコマンドを実行します。実行中は、サーバーサイド

のTypeScriptの変更を感知すると、トランスパイル、サーバーの再起動まで自動で行われます。

リスト3.4.1.1

```
$ cd [ワーキングディレクトリ『chapter-3』のパス]/server
$ npm run dev
```

フロントエンドの開発には、新規でターミナルを開き、リスト3.4.1.2のコマンドを実行します。

リスト3.4.1.2

```
$ cd [ワーキングディレクトリ『chapter-3』のパス]/client
$ npm run start
```

コマンドを実行したら、以下のURLをブラウザで開きます。

```
URL:http://localhost:4200
```

こちらもTypeScriptファイルの変更を監視し、トランスパイルからブラウザのリロードが行われます。

3.4.2本番環境用のスクリプト実行手順

本番環境用では、まずフロントエンドのアプリケーションをビルドします。リスト3.4.2.1のコマンドを実行します。

リスト3.4.2.1

```
$ cd [ワーキングディレクトリ『chapter-3』のパス]/client
$ npm run build
```

フロントエンドのビルドが完了したら、サーバーサイドのアプリケーションを起動します。リスト3.4.2.2のコマンドを実行します。

リスト3.4.2.2

```
$ cd [ワーキングディレクトリ『chapter-3』のパス]/server
$ npm run build
$ npm start
```

ブラウザを起動し、以下のURLにアクセスして、サンプルアプリケーションの画面が表示されていれば完了です。

```
URL: http://localhost:3000/
```

著者紹介

鈴木 潤 (すずき じゅん)

1989年6月27日生まれ。福島県いわき市出身。2013年茨城大学工学部情報工学科を卒業。システムインテグレーターに入社。2016年にHR企業にエンジニアとして転職し今日に至る。好きなマンガはポプテピピック。

◎本書スタッフ
アートディレクター/装丁：岡田章志＋GY
編集協力：飯嶋玲子
デジタル編集：栗原 翔

技術の泉シリーズ・刊行によせて
技術者の知見のアウトプットである技術同人誌は、急速に認知度を高めています。インプレスR&Dは国内最大級の即売会「技術書典」(https://techbookfest.org/) で頒布された技術同人誌を底本とした商業書籍を2016年より刊行し、これらを中心とした『技術書典シリーズ』を展開してきました。2019年4月、より幅広い技術同人誌を対象とし、最新の知見を発信するために『技術の泉シリーズ』へリニューアルしました。今後は「技術書典」をはじめとした各種即売会や、勉強会・LT会などで頒布された技術同人誌を底本とした商業書籍を刊行し、技術同人誌の普及と発展に貢献することを目指します。エンジニアの"知の結晶"である技術同人誌の世界に、より多くの方が触れていただくきっかけになれば幸いです。

株式会社インプレスR&D
技術の泉シリーズ　編集長　山城 敬

●お断り
掲載したURLは2018年3月23日現在のものです。サイトの都合で変更されることがあります。また、電子版ではURLにハイパーリンクを設定していますが、端末やビューアー、リンク先のファイルタイプによっては表示されないことがあります。あらかじめご了承ください。
●本書の内容についてのお問い合わせ先
株式会社インプレスR&D　メール窓口
np-info@impress.co.jp
件名に『本書名』問い合わせ係」と明記してお送りください。
電話やFAX、郵便でのご質問にはお答えできません。返信までには、しばらくお時間をいただく場合があります。なお、本書の範囲を超えるご質問にはお答えしかねますので、あらかじめご了承ください。
また、本書の内容についてはNextPublishingオフィシャルWebサイトにて情報を公開しております。
http://nextpublishing.jp/

●落丁・乱丁本はお手数ですが、インプレスカスタマーセンターまでお送りください。送料弊社負担にてお取り替えさせていただきます。但し、古書店で購入されたものについてはお取り替えできません。
■読者の窓口
インプレスカスタマーセンター
〒101-0051
東京都千代田区神田神保町一丁目105番地
TEL 03-6837-5016／FAX 03-6837-5023
info@impress.co.jp
■書店／販売店のご注文窓口
株式会社インプレス受注センター
TEL 048-449-8040／FAX 048-449-8041

技術の泉シリーズ
TypeScriptで作るシングルページアプリケーション

2018年3月30日　初版発行Ver.1.0（PDF版）
2019年4月5日　Ver.1.1

著　者　鈴木 潤
編集人　山城 敬
発行人　井芹 昌信
発　行　株式会社インプレスR&D
　　　　〒101-0051
　　　　東京都千代田区神田神保町一丁目105番地
　　　　https://nextpublishing.jp/
発　売　株式会社インプレス
　　　　〒101-0051　東京都千代田区神田神保町一丁目105番地

●本書は著作権法上の保護を受けています。本書の一部あるいは全部について株式会社インプレスR&Dから文書による許諾を得ずに、いかなる方法においても無断で複写、複製することは禁じられています。

©2018 Jun Suzuki. All rights reserved.
印刷・製本　京葉流通倉庫株式会社
Printed in Japan

ISBN978-4-8443-9815-8

NextPublishing®

●本書はNextPublishingメソッドによって発行されています。
NextPublishingメソッドは株式会社インプレスR&Dが開発した、電子書籍と印刷書籍を同時発行できるデジタルファースト型の新出版方式です。https://nextpublishing.jp/